LABORATORY RAT

PROCEDURAL TECHNIQUES

MANUAL AND DVD

LABORATORY
RAT

PROCEDURAL TECHNIQUES

MANUAL AND DVD

John J. Bogdanske, BA

Scott Hubbard-Van Stelle, AS, CVT, RLATG

Margaret Rankin Riley, BS

Beth M. Schiffman, BS, CVT, RLATG

CRC Press
Taylor & Francis Group
Boca Raton   London   New York

CRC Press is an imprint of the
Taylor & Francis Group, an **informa** business

CRC Press
Taylor & Francis Group
6000 Broken Sound Parkway NW, Suite 300
Boca Raton, FL 33487-2742

© 2011 by Wisconsin Alumni Research Foundation Laboratory
CRC Press is an imprint of Taylor & Francis Group, an Informa business

No claim to original U.S. Government works

Printed in the United States of America on acid-free paper
10 9 8 7 6 5 4 3 2 1

International Standard Book Number: 978-1-4398-5044-2 (Paperback)

Library of Congress Cataloging-in-Publication Data

Laboratory rat procedural techniques : manual and DVD / John J. Bogdanske ... [et al.].
    p. ; cm.
  Includes bibliographical references.
  ISBN 978-1-4398-5044-2 (pbk. : alk. paper)
    1. Rats as laboratory animals--Handbooks, manuals, etc. I. Bogdanske, John J.
  [DNLM: 1. Animals, Laboratory--Handbooks. 2. Laboratory Techniques and Procedures--Handbooks. 3. Rats--Handbooks. QY 39]

SF407.R38L345 2011
616'.02733--dc22
                                             2010039395

Visit the Taylor & Francis Web site at
http://www.taylorandfrancis.com

and the CRC Press Web site at
http://www.crcpress.com

contents

Section I
rat DVD text and voice-over

Section II
rat procedural technique handouts

Supplementary Resources Disclaimer

Additional resources were previously made available for this title on DVD. However, as DVD has become a less accessible format, all resources have been moved to a more convenient online download option.

You can find these resources available here: http://resourcecentre.routledge.com/books/9781439850442

Please note: Where this title mentions the associated disc, please use the downloadable resources instead.

disclaimers

This DVD is not intended for use, and must not be used, by individuals who lack thorough training and understanding of animal biosafety. Rather, this video should be used as a resource or refresher for techniques previously learned through animal biosafety training.

The techniques depicted in this video must be performed in compliance with institutionally approved animal protocols, observing all institutional and governmental safety regulations with regard to the use of appropriate safety equipment and procedures. If you lack a thorough understanding of the institutional and governmental rules applicable to the techniques depicted in this video, you must not perform these techniques.

The techniques depicted in this video must be performed only in an approved animal facility under the guidance of animal care specialists.

The techniques depicted in this video take place in a controlled laboratory setting and present a risk of serious bodily injury (including death) through the following exposure hazards: sharp objects; chemical or pharmaceutical agents administered to rats; and biohazards, including but not limited to the bodily fluids and waste of rats. These hazards can be avoided only by following a thorough laboratory safety program that complies with all applicable institutional and governmental rules. If your facility lacks such a program or if you do not thoroughly understand all aspects of that program, you must not perform the techniques depicted in this video.

introduction

The trainers at the Research Animal Resources Center (RARC) at the University of Wisconsin–Madison thank you for taking the time to view this training DVD. This DVD should be viewed as a resource or a refresher for techniques previously learned in an animal biosafety class or in your own research setting. The procedures and techniques demonstrated are approved by the University of Wisconsin Institutional Animal Care and Use Committee (IACUC) and are commonly used on research rats. If after viewing this DVD you find that a specific method or technique is not covered in enough detail, or that you would prefer further instruction, please feel free to contact a trainer via e-mail at trainer@rarc.wisc.edu, and we will be happy to assist you. It is our hope that you find this information both educational and useful.

1

rat DVD text and voice-over

disease management

Surveillance, diagnosis, treatment, and control of disease are integral components of adequate veterinary care.[1] Subclinical microbial, particularly viral, infections can occur in barrier and conventionally maintained animals. Such infections can seriously compromise experimental protocols by inducing profound changes in immunologic, physiologic, neoplastic, and toxicologic responses in infected animals. Therefore, control and elimination of known pathogens is vital for good science as well as the health and well-being of research animals.

Please note that in the following demonstrations, the technicians are performing the techniques wearing exam gloves and a lab coat. Check the standard operating procedures (SOPs) of your laboratory or the requirements of your animal facility and ensure that the proper personal protective equipment is worn while working with animals to prevent possible injury to yourself and the spread of disease.

---

[1] American Association for Laboratory Animal Science. *Laboratory Mouse Handbook*. Memphis, TN: American Association for Laboratory Animal Science, 2006.

rat handling/transfer

Safely transferring animals from cage to cage or cage to work area is the first task you need to perform effectively when working with rodents. The demonstrations show various methods of transferring rats from a shoebox cage and subsequent restraint. Watch closely as the technician picks up the rat by the base of the tail. This is the best method to prevent the possibility of stripping the skin off the rat's tail.

Voice-Over: Safely transferring rats from cage to cage is accomplished easily by picking up a rat by the base of the tail from one cage and securely moving it to the other.

rat handling/restraint

The demonstration shows you how to restrain rats using one hand. Watch closely as the technician holds the rat by the base of the tail and slides the index and middle fingers over the shoulder of the rat while grasping its body with the thumb and ring finger. This method of restraint is commonly used for a number of procedures.

Voice-Over 1: Grasp the base of the rat's tail with your non-dominant hand. Lift and move the rat to your forearm while switching the tail hold to the dominant hand. Slide your index and middle fingers over the rat's shoulders while wrapping the thumb and ring finger around the abdomen to secure the restraint. Mastering this technique will allow you to safely perform a number of procedures.

Voice-Over 2: As your handling skills improve, you might find it easier to remove rats from their cages by first securing the rat's tail by the base with your dominant hand while the rat is inside the cage and then sliding the index and middle fingers of your nondominant hand over the rat's shoulder, wrapping the thumb and ring finger around the abdomen as shown. You can now continue with your procedure.

one-handed injection technique

There are injection techniques that require the ability to manipulate a syringe using only one hand. A brief demonstration of this technique is given.

Voice-Over: There are a number of techniques that require the ability to manipulate a syringe using only one hand. Stabilize the syringe with your thumb and index and middle fingers while using the ring finger to work the plunger. You can easily practice this in your lab.

intraperitoneal (IP) injection
(one person)

Use the one-person intraperitoneal (IP) injection method to inject an anesthetic agent or medication (10 ml/kg maximum). Use caution when holding and injecting as damage to internal organs can occur if care is not taken to properly restrain the animal and control the insertion of the needle. Experience performing one-handed injections is necessary for this technique.

Have the needle and syringe ready and loaded with the solution to be injected. Restrain the rat using your nondominant hand. Divide the abdomen into quadrants. You will inject into the lower right quadrant of the animal. Carefully insert the needle at a 45° angle into the abdomen until you sense the needle "pop" through the skin and abdominal wall. Draw back on the plunger. If you observe a vacuum bubble, inject your solution. When finished, dispose of the needle and syringe in a sharps container. If while drawing back you get a colored liquid, pull the needle out; discard the needle, syringe, and solution into a sharps container; and start over.

recommended supplies

- 1- or 3-ml syringe
- 21- to 25-gauge needle

Voice-Over 1: This illustration shows the rat abdomen divided into quadrants. It is recommended that IP injections be given into the lower right quadrant as indicated by the red dot. This area, cranial to the knee and left of the midline, will allow you to avoid accidentally injecting into the testicles of the male rat if they are retracted into the abdomen.

Voice-Over 2: Use the restraint techniques described in Chapter 2 to place the rat in a position that will allow you to view the abdominal area.

Voice-Over 3: Once proper restraint is achieved, insert the needle through the skin and abdominal wall into the animal's lower right quadrant. Draw back on the plunger. If you observe a vacuum bubble, proceed to inject your solution. If while drawing back you get a colored liquid, pull the needle out; discard the needle, syringe, and solution into a sharps container; and start over.

intraperitoneal (IP) injection
(two person)

Use the two-person intraperitoneal (IP) injection method to inject an anesthetic agent or medication (10 ml/kg maximum). Use caution when holding and injecting as damage to internal organs can occur if care is not taken to properly restrain the animal and control the insertion of the needle.

Have the needle and syringe already loaded with the solution to be injected. Restrain the head area of the rat with one hand while holding the tail with the other. Divide the abdomen into quadrants and have your partner inject into the lower right quadrant of the animal. Carefully insert the needle at a 45° angle into the abdomen until the needle "pops" through the skin and abdominal wall. Draw back on the plunger. When you observe a vacuum bubble, inject your solution. When finished, dispose of the needle and syringe in a sharps container. If while drawing back you get a colored liquid, pull the needle out; discard the needle, syringe, and solution into a sharps container; and start over.

recommended supplies

- 1- or 3-ml syringe
- 25-gauge needle

Voice-Over: In some cases, it may be helpful to have a lab partner restrain the rat for you. Once restraint is achieved, stand to the left of your partner, grasp the rat's right hind leg, and move aside. Administer an IP injection as previously described.

intraperitoneal (IP) injection
(one-person towel method)

Use the one-person towel method to inject an anesthetic agent or medication (10 ml/kg maximum) via the intraperitoneal (IP) route. Use caution when holding and injecting as damage to internal organs can occur if care is not taken to properly restrain the animal and control the insertion of the needle. Experience performing one-handed injections is necessary for this technique.

Use a sturdy cloth towel opened on a table. Place the rat at the center of the towel and fold the towel over the rat with your nondominant hand, leaving the tail area exposed. Grasp the right hind leg and roll the rat over slightly to expose the abdomen. Insert the needle, angled slightly toward the head, into the lower right quadrant of the animal. You will sense the needle "pop" through the skin and abdominal wall. Draw back on the plunger. When you observe a vacuum bubble, inject your solution. When done, pull the needle out and dispose of it in a sharps container. If while drawing back you get a colored liquid, pull the needle out; discard the needle, syringe, and solution into a sharps container; and start over.

recommended supplies

- 1- or 3-ml syringe
- 25-gauge needle
- Towel

Voice-Over 1: The restraint is the only difference between the standard IP injection technique and the towel method. Keeping the rat under the towel may be the biggest challenge for this method. Insert the needle into the animal's lower right quadrant and draw back on the plunger; if you observe a vacuum, proceed to inject your solution. If while drawing back you get a colored liquid, pull the needle out; discard the needle, syringe, and solution into a sharps container; and start over.

Voice-Over 2: Restraining the rat with the right hand may allow for a more favorable view of the rat's lower right quadrant as demonstrated in this clip. Complete the technique by inserting the needle, drawing back on the plunger to observe a vacuum, and injecting your solution. As previously stated, if while drawing back you get a colored liquid, pull the needle out; discard the needle, syringe, and solution into a sharps container; and start over.

subcutaneous (SQ) injection
(one-person towel method)

For the one-person towel method of subcutaneous (SQ) injection, use a sturdy cloth towel opened on a table. Place the rat at the center of the towel and fold the towel over the rat, leaving the tail area exposed. Restrain the front two-thirds of the rat with the palm of your nondominant hand while tenting the skin over the hindquarters with your thumb and index finger. Insert the needle at the base of the tented folds of skin. Pull back on the plunger of the syringe to verify that a vacuum is created and inject the solution. In general, no greater than 5 ml should be injected per subcutaneous injection site (300-g adult rat). Use several sites on the back of the animal if larger volumes must be administered. Experience performing one-handed injections is necessary for this technique.

recommended supplies

- 1- or 3-ml syringe
- 22-gauge needle
- Towel

Voice-Over 1: Place the rat at the center of a towel and fold the towel over the rat, leaving the tail area exposed.

Voice-Over 2: Restrain the front two-thirds of the rat with the palm of your nondominant hand while tenting the skin over the hindquarters with your thumb and index finger. Insert the needle at the base of the tented folds of skin. Draw back to verify that a vacuum is created and proceed to inject the solution.

pedal vein blood draw

The pedal vein blood draw technique is used to collect relatively small volumes of blood, up to 100 µl. See the Appendix on page 79 for information on calculating the blood volume and maximum blood sample volume of the rat. The trainers at the Research Animal Resources Center (RARC) recommend using 10% of the total blood volume as the amount that can be collected at one time. To allow for an adequate recovery period, we further recommend that the maximum amount be withdrawn only once every 2 weeks (based on 10%). Daily samples can be collected as long as the cumulative blood volume does not exceed the calculated maximum over a 7-day period.

Place a tourniquet as high up on the rear leg of the rat as possible. Apply a drop of artificial tears (or petroleum jelly) near the top of the foot and rub across the surface. The ointment will allow the blood to bead up on the skin. Using a 25-gauge needle, gently poke into the vein, taking care not to pierce completely through it because this will result in blood accumulating under the skin. As the blood beads up, use a heparinized capillary tube to collect it. When finished, release the tourniquet and apply pressure with gauze to stop the bleeding.

recommended supplies

- Heparinized capillary tube
- Artificial tears or petroleum jelly
- Tourniquet
- 25-gauge needle
- Gauze

Voice-Over 1: Place a tourniquet as high up on the rat's rear leg as possible.

Voice-Over 2: Apply petroleum jelly near the top of the foot and rub across the surface. The ointment will allow the blood to bead up on the skin.

Voice-Over 3: Using a 25-gauge needle, gently poke into the vein, being careful not to pierce completely through it as this will result in blood accumulating under the skin. As the blood beads up, use a heparinized capillary tube to collect the blood.

Voice-Over 4: When finished, release the tourniquet and apply pressure with gauze to stop the bleeding.

saphenous vein blood draw

The saphenous vein blood draw technique is used to collect relatively small volumes of blood, less than 50 μl. See the Appendix on page 79 for information on calculating the blood volume and maximum blood sample volume for the rat. The trainers at the Research Animal Resources Center (RARC) recommend using 10% of the total blood volume as the amount that can be collected at one time. To allow for an adequate recovery period, we further recommend that the maximum amount be withdrawn only once every 2 weeks (based on 10%). Daily samples can be collected as long as the cumulative blood volume does not exceed the calculated maximum over a 7-day period.

Clip the lateral surface of the hind leg of the rat. Place a tourniquet as high up on the rear leg of the rat as possible. Apply a drop of artificial tears (or petroleum jelly) on the lateral side of the leg and rub across the surface. The ointment will allow the blood to bead up on the skin. Using a 27-gauge needle, gently poke into the vein, taking care not to pierce completely through it because this will result in blood accumulating under the skin. As the blood beads up, use a heparinized capillary tube to collect it. When finished, release the tourniquet and apply pressure with gauze to stop the bleeding.

recommended supplies

- Heparinized capillary tube
- Clippers
- Artificial tears or petroleum jelly
- Tourniquet
- 27-gauge needle
- Gauze

Voice-Over 1: Clip the lateral surface of the rat's hind leg to expose the saphenous vein.

Voice-Over 2: Position a tourniquet as high up on the rat's hind leg as possible.

Voice-Over 3: Apply artificial tears over the vein and spread evenly. The ointment will allow the blood to bead up on the skin.

Voice-Over 4: Using a 27-gauge needle, gently poke into the vein, being careful to not pierce completely through it as this will result in blood accumulating under the skin. As the blood beads up, use a heparinized capillary tube to collect the blood. When finished, release the tourniquet and apply pressure with gauze to stop the bleeding.

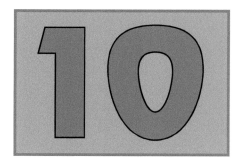

oral gavage

Oral gavage in the rat is used to administer liquid food, experimental solutions, or medication directly to the stomach. The recommended normal volume is 10 ml/kg.

Restraint is the key to this technique. You must have the ability to properly restrain a rat to successfully administer the solution. The length of the feeding needle should be the same as the distance from the nose to the stomach (just below rib cage) of the rat. Once proper restraint is achieved, place the feeding needle in the back of the mouth of the rat and tilt its head back until the syringe is parallel with the body of the rat. The needle will slide down easily when positioned properly. *Do not use force.* It is possible to perforate the esophagus or force the needle into the trachea. Proper insertion will be achieved using gravity and the weight of the syringe. If there is any resistance, *stop,* pull the needle out, and try again. Insert the feeding needle until only the needle hub is showing. At this time, inject your solution. Slowly remove the needle following its curve as you pull it out.

recommended supplies

- 1- or 3-ml Luer lock syringe
- 18-gauge, 3-inch feeding needle (based on a 300-g adult rat)

Voice-Over 1: Once proper restraint is achieved, determine if the length of the needle is appropriate by measuring the distance from the rat's nose to its stomach.

Voice-Over 2: Place the feeding needle in the back of the mouth and tilt the head back until the syringe is parallel with the rat's body. The needle will slide down easily when positioned properly. Insert the needle until only the hub is showing; you can now safely inject your solution. *Remember, do not use force.* Proper insertion will be achieved using gravity and the weight of the syringe.

Voice-Over 3: In this clip, the oral gavage technique is performed by the technician while holding the rat against his/her body.

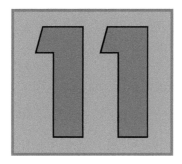

jugular bleed (two person)

The jugular bleed is used to collect relatively large volumes of blood. See the Appendix on page 79 for information on calculating the blood volume and maximum blood sample volume of the rat. The trainers at the Research Animal Resources Center (RARC) recommend using 10% of the total blood volume as the amount that can be collected at one time. To allow for an adequate recovery period, we further recommend that the maximum amount be withdrawn only once every 2 weeks (based on 10%). Daily samples can be collected as long as the cumulative blood volume does not exceed the calculated maximum over a 7-day period.

You will need two people to perform this technique, which uses a bleeding board as shown in the video clip. Proper restraint is required to be successful. Once the rat is restrained, find your landmarks for the needle insertion into the jugular vein. The head cone is used to rotate the head of the rat away from the bleeding site. The jugular vein is very superficial, and you will need to insert the needle slowly, drawing back on the plunger at the same time. Stop when you get a flash of blood in the hub, then draw back on the plunger to obtain the sample. To prevent injury to the rat *do not* insert the needle more than half of its length. When finished and the rat is released from the bleeding board, be sure that the bleeding has stopped before placing the rat back in its cage.

recommended supplies

- 3-ml syringe
- Rat bleeding board
- 23-gauge needle
- Head cone restraint

Voice-Over 1: Proper restraint as demonstrated in the first part of this clip is necessary to perform this technique successfully.

Voice-Over 2: Cover the rat's head with the restraint device as shown, then rotate the head and move it away from the bleeding site, being careful not to occlude the rat's airway.

Voice-Over 3: In this clip, you can see the triangle shaped area that is formed after proper positioning is achieved. The red dot indicates where the needle needs to be inserted.

Voice-Over 4: Knowing the jugular vein is very superficial, you will not need to insert the needle more than half of its length. Insert the needle while drawing back on the plunger; as soon as there is a flash of blood in the needle hub, stop inserting and draw back on the plunger to obtain the sample.

Voice-Over 5: Care is taken to secure the rat before releasing the string leg restraints. The technician is able to pinch the bleeding site with his/her index finger and thumb while restraining the rat to ensure the bleeding has stopped.

tail artery blood draw

Use the tail artery blood draw technique to collect relatively large volumes of blood. See the Appendix on page 79 for information on calculating the blood volume and maximum blood sample volume of the rat. The trainers at the Research Animal Resources Center (RARC) recommend using 10% of the total blood volume as the amount that can be collected at one time. To allow for an adequate recovery period, we further recommend that the maximum amount be withdrawn only once every 2 weeks (based on 10%). Daily samples can be collected as long as the cumulative blood volume does not exceed the calculated maximum over a 7-day period.

Begin by warming a gel pack in a microwave. Touch the pack to your wrist to ensure that it is not too hot before wrapping it around the tail of the rat. Warming the tail will cause the artery to dilate. Use of isoflurane anesthesia will also increase vasodilation. The tail artery is on the ventral surface of the tail. Start one-third from the distal end of the tail and work toward the base in the event the initial try is unsuccessful. Insert the needle bevel side up into the artery while drawing back on the plunger. A flash of blood will appear in the needle hub when you are in the artery. Slowly draw back on the plunger until the desired volume is collected. If the artery bleeds after the needle is removed, apply slight pressure to the area with gauze.

recommended supplies

- 3-ml syringe
- 23-gauge needle
- Gel pack

Voice-Over 1: Begin by warming a gel pack in a microwave for approximately 30 seconds. Touch the pack to your wrist to ensure it is not too hot before wrapping it around the rat's tail; warming the tail will cause the tail artery to dilate. Start one-third from the distal end of the tail and work toward the base in the event that the initial try is unsuccessful. Insert the needle, bevel side up, into the artery while drawing back on the plunger. You will get a flash of blood into the needle hub when you are in the artery. Draw back on the plunger while keeping the needle steady until you obtain the appropriate sized blood sample.

Voice-Over 2: When finished, apply pressure with gauze to ensure the bleeding has stopped.

tail vein injection

Tail veins are one of the few intravenous injection sites on the rat. Proficiency in one-handed injections is necessary. The normal bolus volume is 5 ml/kg.

The veins are located superficially on the lateral sides of the tail. To dilate the tail veins, begin by warming the tail with a warm gel pack or recirculating water blanket. Use of isoflurane anesthesia will also cause the veins to dilate. Start one-third from the distal end of the tail and work toward the base in the event that the initial try is unsuccessful. Insert the needle bevel side up into the vein while drawing back on the plunger. You will get a flash of blood into the needle hub when you are in the vein. Begin injecting; if the blood proximal to the needle vacates the vein, continue to slowly inject. There will be virtually no resistance when correctly injecting into the vein. If a whitish bleb or bubble appears under the skin, stop injecting, remove the needle, and choose a site closer to the base of the tail.

recommended supplies

- 3-ml syringe
- 25-gauge needle
- Gel pack
- Gauze

Voice-Over 1: Begin by warming a gel pack in a microwave for approximately 30 seconds. Touch the pack to your wrist to ensure it is not too hot before wrapping it around the rat's tail; warming the tail will cause the veins to dilate.

Voice-Over 2: This illustration shows the proper position of the needle in relation to the rat's tail vein. Taking the time to align the two will increase the probability of success.

Voice-Over 3: Start one-third from the distal end of the tail and work toward the base in the event that the initial try is unsuccessful.

Voice-Over 4: Insert the needle bevel side up into the vein while drawing back on the plunger. You will get a flash of blood into the needle hub when you are in the vein. Begin injecting; if the blood proximal to the needle vacates the vein, continue to slowly inject. There will be virtually no resistance when correctly injecting into the vein. If a whitish bleb or bubble appears under the skin, stop injecting, remove the needle, and choose a site closer to the base of the tail.

Voice-Over 5: When finished, use gauze and apply slight pressure to stop the bleeding.

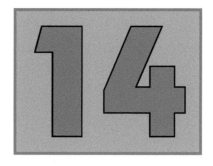

ear notching

Ear notching/punching is a long-term identification method that can be performed quickly with little pain or distress. Using an ear punch, holes or notches are placed in the ears following a simple identification chart. Removed tissue can be used for genotyping, possibly replacing the need for a tail biopsy. Care must be taken when placing the punches; if tags are too close to the edge of the ear, rats may rip or tear open the punches, leaving patterns hard to identify.

recommended supplies

- Scissors-style or thumb-style ear punch

Voice-Over 1: (Pic. #1) This illustration shows an identification code that can be used to identify rodents with numbers 1 through 399.

Voice-Over 2: (Pic. #2) Shown from top to bottom: ear tag applicator, ear tags, thumb- and scissor-style ear punches.

Voice-Over 3: Punching or notching holes at various positions in the ears requires the use of an ear punch. Grasp the rat and, while following the code, place the notches or punches into the appropriate ear.

Voice-Over 4: Check the markings regularly to ensure the rats can be identified accurately, especially if rats are group housed. During fights or grooming, the rats could rip each other's ears, making identification impossible.

Voice-Over 5: Removed tissue can be used for genotyping, possibly replacing the need for a tail biopsy.

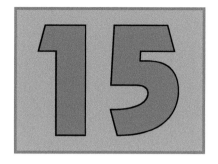

ear tagging

Ear tagging is a long-term identification method that can be performed quickly with little pain or distress. For demonstration purposes, the rat in the following clip was anesthetized, and a black dot was placed on the ear to indicate the preferred location for the ear tag.

Using an ear tag applicator, a uniquely numbered tag is placed in the lower one-third of the ear of the rat. Use care when placing the tags; if tags are placed too close to the edge of the ear, rats can tear the tag out.

recommended supplies

- Numbered tags
- Tag applicator

Voice-Over 1: This clip shows how to correctly place an ear tag into the applicator tip, followed by what the tag looks like after it has been crimped.

Voice-Over 2: Position the tag applicator over the lower third of the rat's ear and crimp. Ideally, the number faces forward for easy identification.

II

rat procedural technique handouts

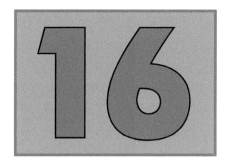

intraperitoneal (IP) injection

Purpose (what you will learn): How to safely administer an intraperitoneal (IP) injection.

Use this technique if: Using large-volume liquids (10 ml/kg maximum volume), for example, if you are administering drug doses that have been diluted. Injection volume information can be found at http://www.rarc.wisc.edu.

Recommended skills: Basic handling and restraint of a rat. Use caution when holding and injecting because damage to internal organs could occur if care is not taken to restrain the animal and control the insertion of the needle.

Recommended supplies: A 21- to 25-gauge needle, 1- or 3-ml syringe, exam gloves, and sharps container.

You must be able to adequately restrain a rat before performing any injection technique.

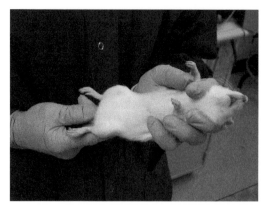

By turning the rat over, you can view the abdomen. Make sure its chest is moving up and down and that you are not restricting breathing.

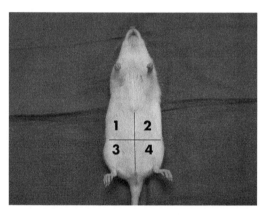

Quadrant 3 is the best area for the IP injection. Inject halfway between the midline and where the leg attaches to the body.

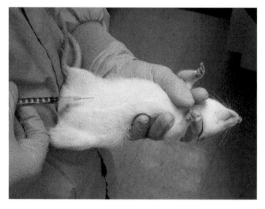

The needle should be at a 45° angle. You will sense the needle pop through the skin and abdominal wall. Draw back slightly on the plunger. When you observe a vacuum bubble, inject your solution.

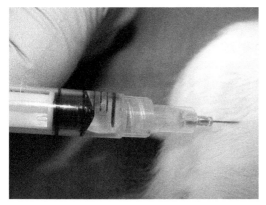

If while drawing back you get a colored liquid, pull the needle out, discard the needle and syringe, and start over with a new solution. Always dispose of the needle and syringe into a sharps container.

subcutaneous (SQ) injection: towel method

Purpose (what you will learn): How to safely administer a subcutaneous (SQ) injection.

Use this technique if: Injecting up to 5 ml/kg (maximum is 10 ml/kg) of liquid. Use several sites over the back of the animal if larger volumes must be administered. Injection volume information can be found at http://www.rarc.wisc.edu.

Recommended skills: Basic handling and restraint of a rat.

Recommended supplies: A 22-gauge needle, 1- or 3-ml syringe, towel, exam gloves, and sharps container.

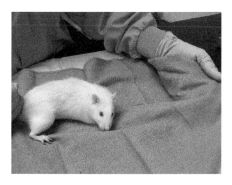

You must be able to adequately restrain a rat before performing any injection technique. The one-person towel method is used for this demonstration.

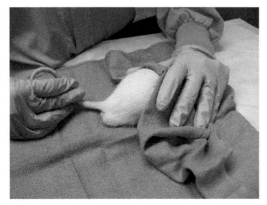

Place the rat at the center of the towel and proceed to fold the towel over the rat, leaving the tail area exposed.

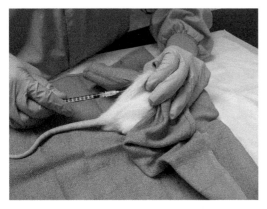

Restrain the front two-thirds of the rat with the palm of your nondominant hand while tenting the skin over the hindquarters with your thumb and index finger.

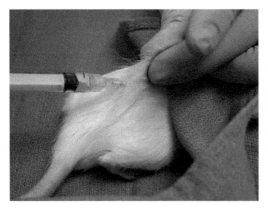

Insert the needle at the base of the tented folds of skin. Pull back on the plunger of the syringe to verify that a vacuum is created and inject the solution.

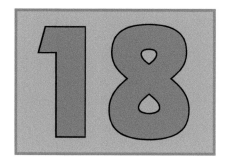

jugular bleed

Purpose (what you will learn): How to safely draw blood from the jugular vein of the rat. This is a two-person technique if anesthesia is not used.

Use this technique if: Collecting relatively large volumes of blood. See the Appendix on page 79 for information on calculating the blood volume and maximum blood sample volume of the rat. The trainers at the Research Animal Resources Center (RARC) recommend using 10% of the total blood volume as the maximum amount that can be collected at one time. To allow for an adequate recovery period, we further recommend that the maximum amount be withdrawn only once every 2 weeks (based on 10%). Daily samples can be collected as long as the cumulative blood volume does not exceed the calculated maximum over a 7-day period.

Recommended skills: You must know and understand the proper method to restrain a rat for this technique. The ability to operate a syringe and needle with one hand is necessary. Knowledge and experience of using an anesthesia machine are also helpful.

Recommended supplies: A 23-gauge needle, 3-ml syringe, rat bleeding board, head cone for restraint, exam gloves, and sharps container. (Anesthesia machine may be used if listed in animal protocol.)

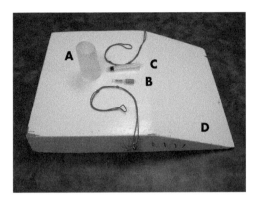

Recommended supplies for the jugular bleed:

A. Head cone restraint

B. 23-gauge needle

C. 3-ml syringe

D. Bleeding board

The pictures below show the proper restraint of an unanesthetized rat placed on the bleeding board.

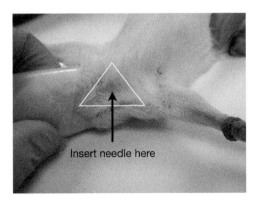

Find your landmarks for the needle insertion into the jugular vein. With the head cone in place, rotate the head (in this example, clockwise), then move the rotated head away from the bleeding site.

Insert needle here

Right external jugular

Left internal jugular

Right axillary

The jugular vein is very superficial; insert the needle while slowly drawing back on the plunger. Stop insertion when you see a flash of blood in the hub of the needle; draw back on the plunger to obtain a sample. To prevent injury to the rat do not insert the needle more than half of its length.

Ensure that the jugular vein is no longer bleeding before returning the rat to its cage. Pinching the bleeding area between the thumb and index finger will stop any bleeding.

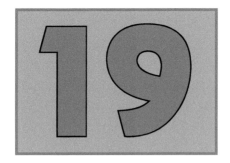

oral gavage

Purpose (what you will learn): How to safely feed or dose by mouth using a feeding needle.

Use this technique if: You need to administer liquid food, experimental solutions, or medication directly to the stomach of the rat. The recommended normal volume is 10 ml/kg.

Recommended skills: Basic handling and restraint of a rat while inserting a gavage or feeding needle into the rat esophagus.

Recommended supplies: An 18-gauge, 3-inch feeding needle; 1- or 3-ml Luer lock syringe; exam gloves; and sharps container.

You must be able to adequately restrain a rat before proceeding with this technique.

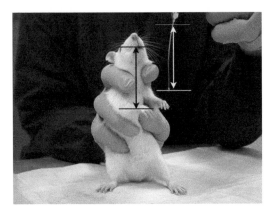

Once the proper restraint is achieved, determine if the length of the needle is appropriate. The length of the needle should be the same as the distance from the nose of the rat to its stomach.

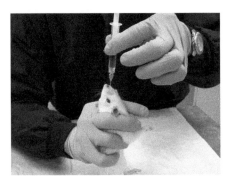 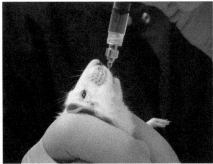

Place the feeding needle in the back of the mouth of the rat and tilt the head back until the head is parallel with its body as shown in the picture on the left. The needle will slide down easily when positioned properly. Do not use force. It is possible to perforate the esophagus or force the needle into the trachea. Proper insertion will be achieved using gravity and the weight of the syringe. If there is any resistance, stop, pull the needle out, and try again. When the technique is done correctly, you will not see the gavage needle, only the hub as shown at right. At this time, inject your solution. Slowly remove the needle, following its curve as you pull it out.

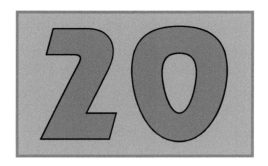

pedal vein blood draw

Purpose (what you will learn): How to safely draw blood from the pedal vein of the rat. You will need a restraint device if anesthesia is not used.

Use this technique if: Collecting relatively small volumes of blood, up to 100 µl. See the Appendix on page 79 for information on calculating the blood volume and maximum blood sample volume of the rat. The trainers at the Research Animal Resources Center (RARC) recommend using 10% of the total blood volume as the maximum amount that can be collected at one time. To allow for an adequate recovery period, we further recommend that the maximum amount be withdrawn only once every 2 weeks (based on 10%). Daily samples can be collected as long as the cumulative blood volume does not exceed the calculated maximum over a 7-day period.

Recommended skills: Basic handling and restraint of a rat. Knowledge and experience of using an anesthesia machine are also helpful.

Recommended supplies: A 25- to 27-gauge needle, tourniquet, petroleum jelly or artificial tears, blood collection tube, heparinized capillary tube (inside diameter 1.1 mm, wall 0.2 × 75 mm L) exam gloves, anesthesia machine, and sharps container.

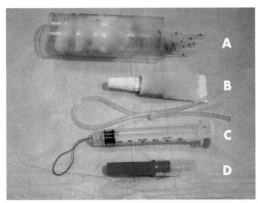

Recommended supplies for the pedal vein blood draw:

A. Heparinized capillary tube
B. Artificial tears
C. Tourniquet
D. 25- to 27-gauge needle

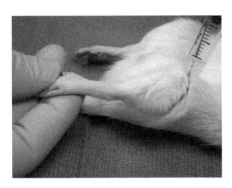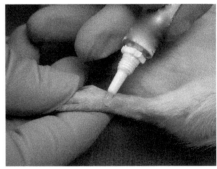

Left: Place a tourniquet as high up on the leg of the rat as possible. Right: Apply the artificial tears (or petroleum jelly) to the leg as shown. Rub the artificial tears around the top of the foot. The ointment will allow the blood to bead up on the skin surface instead of spreading across the foot.

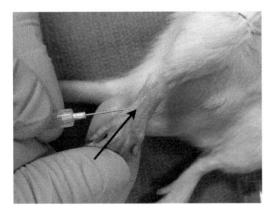

Going in the direction of the arrow, gently poke into the vein, taking care not to pierce completely through because this will result in blood accumulating under the skin.

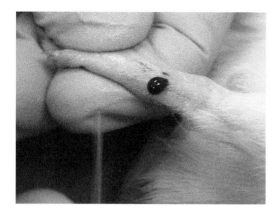

As the blood beads up on the foot, use the heparinized capillary tube to collect the blood.

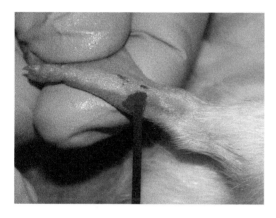

When finished, release the tourniquet and apply pressure with gauze to stop the bleeding.

saphenous vein blood draw

Purpose (what you will learn): How to safely obtain a blood sample from the saphenous vein of the rat. You will need a restraint device if anesthesia is not used.

Use this technique if: Collecting relatively small volumes of blood. See the Appendix on page 79 for information on calculating the blood volume and maximum blood sample volume of the rat. The trainers at the Research Animal Resources Center (RARC) recommend using 10% of the total blood volume as the maximum amount that can be collected at one time. To allow for an adequate recovery period, we further recommend that the maximum amount be withdrawn only once every 2 weeks (based on 10%). Daily samples can be collected as long as the cumulative blood volume does not exceed the calculated maximum over a 7-day period.

Recommended skills: Basic handling and restraint of a rat. Know how to properly use an anesthesia machine if rats will be anesthetized.

Recommended supplies: Small clippers, 25- to 27-gauge needle, heparinized capillary tube (inside diameter 1.1 mm, wall 0.2 × 75 mm L), blood collection tube, exam gloves, 2 × 2 gauze, petroleum jelly or artificial tears, anesthesia machine, and a tourniquet.

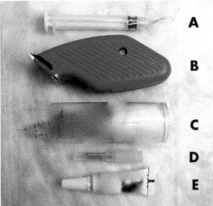

Recommended supplies for the saphenous vein blood draw:

A. Tourniquet

B. Clippers

C. Heparinized capillary tube

D. 25- to 27-gauge needle

E. Artificial tears

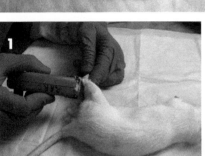

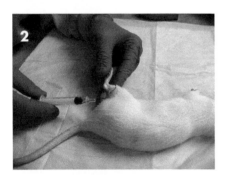

Clip the lateral surface of the hind leg of the rat (1). Place a tourniquet as high up on the hind leg as possible (2). Apply the artificial tears to the leg as shown (3) and spread on the lateral surface. The ointment allows the blood to bead up on the skin surface instead of spreading across the leg.

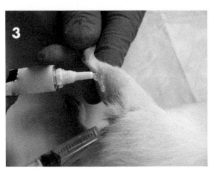

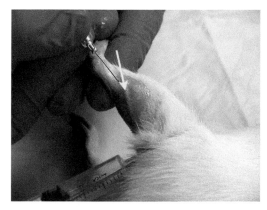

Moving in the direction of the arrow, gently poke into the vein, taking care not to pierce completely through because this will cause blood to collect under the skin.

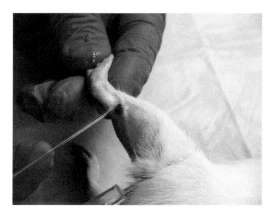

As the blood beads up on the leg, use the heparinized capillary tube to collect the blood.

When finished, release the tourniquet and apply momentary pressure with gauze to stop the bleeding.

tail artery blood draw

Purpose (what you will learn): How to safely draw blood from the tail artery of the rat. You will need a restraint device if anesthesia is not used.

Use this technique if: Collecting relatively large volumes of blood. See the Appendix on page 79 for information on calculating the blood volume and maximum blood sample volume of the rat. The trainers at the Research Animal Resources Center (RARC) recommend using 10% of the total blood volume as the maximum amount that can be collected at one time. To allow for an adequate recovery period, we further recommend that the maximum amount be withdrawn only once every 2 weeks (based on 10%). Daily samples can be collected as long as the cumulative blood volume does not exceed the calculated maximum over a 7-day period.

Recommended skills: Basic handling and restraint of a rat. Knowledge and experience of using an anesthesia machine are also helpful.

Recommended supplies: A 23-gauge needle, 3-ml syringe, gel pack or heated water blanket, exam gloves, anesthesia machine, and sharps container.

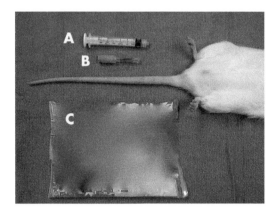

Recommended supplies for the tail artery blood draw:

A. 3-ml syringe
B. 23-gauge needle
C. Gel pack

The picture shows a rat being prepared for a blood draw. In this case, the rat is anesthetized, and a warm gel pack is placed around the tail.

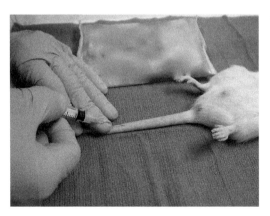

The tail artery will be visible on the ventral side of the tail. Veins are located superficially on the lateral sides of the tail at the 10 and 2 o'clock positions.

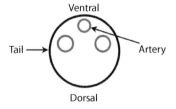

Visualize the artery and insert the needle while applying pressure distally on the tail.

When the blood flashes into the syringe, stop advancing the needle and slowly draw back on the plunger to obtain the blood sample.

If the artery bleeds after the needle is removed, apply slight pressure to the area using gauze.

tail vein injection

Purpose (what you will learn): How to safely inject the tail vein of the rat. You will need a restraint device if anesthesia is not used.

Use this technique if: Injecting solutions into the rat intravenously. Bolus volume is 5 ml/kg; slow injection is 20 ml/kg.

Recommended skills: Basic handling and restraint of a rat. Knowledge and experience of using an anesthesia machine are also helpful.

Recommended supplies: 25-gauge needle, 3-ml syringe, gel pack or heated water blanket, exam, gloves, gauze, anesthesia machine, and sharps container.

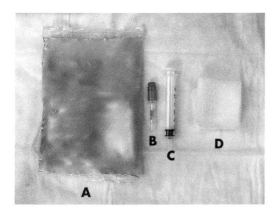

Recommended supplies for the tail vein injection:

A. Gel pack

B. 25-gauge needle

C. 3-ml syringe

D. Gauze

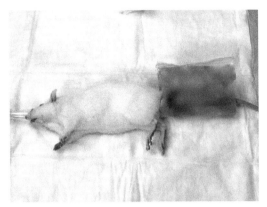

The picture shows a rat being prepared for a tail vein injection. In this case, the rat is anesthetized, and a warm gel pack is placed around the tail to dilate the blood vessels. If isoflurane gas anesthesia is used, it will also cause the vein to vasodilate.

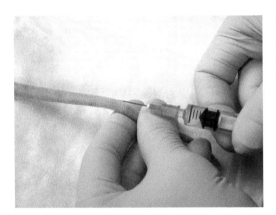

The tail veins are located superficially on the lateral sides of the tail at the 10 and 2 o'clock positions.

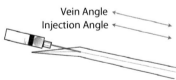

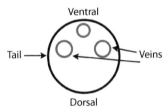

Start at the most distal end of the tail and work toward the base in the event that the initial try is unsuccessful.

Insert the needle into the vein, draw back slightly on the plunger, and look for blood to flash back into the needle hub. Begin injection; if the blood proximal to the needle vacates the vein, continue to inject. There will be virtually no resistance when correctly injecting into the vein. If a bubble of solution appears under the skin, stop injecting, remove the needle, and choose a site closer to the base of the tail.

The tail vein injection, like its counterpart the tail artery draw, can be frustrating for the beginner. Practice and using the appropriate equipment will make the procedure easier. Be sure to take advantage of a heat source as mentioned because this will greatly increase your chances of identifying the vein.

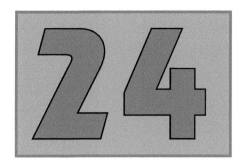

ear notching/punching and ear tags

Purpose (what you will learn): How to safely perform ear notching/punching as well as placing ear tags in rodents.

Use this technique if: You need to individually identify mice or rats.

Recommended skills: You must know and understand the proper methods to restrain a mouse or rat for this technique.

Recommended supplies: Thumb- or scissor-style ear punch or ear tag applicator and numbered ear tags.

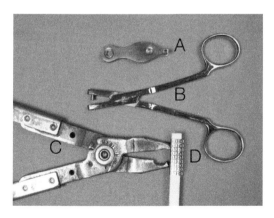

Recommended supplies for ear notching or ear tagging:

A. Thumb-style ear punch

B. Scissor-style ear punch

C. Ear tag applicator

D. Numbered ear tags

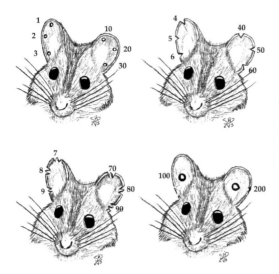

This illustration shows a rodent identification code that can be used to identify rodents with numbers 1 through 399.

ear notching/punching

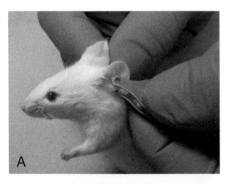

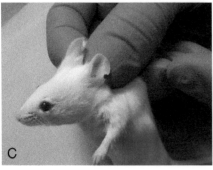

Ear notching/punching is a long-term identification method that can be performed quickly with little pain or distress. In this series of pictures, the mouse is restrained by the scruff (A), and using an ear punch, holes (B) or notches (C) are placed in the ears following the previously shown chart. Care must be taken when placing the punches.

Consistency in placing the notches/punches is necessary to allow for easy reading. Punches too close to the edge of the ear may be ripped or torn open, leaving patterns hard to identify.

ear tagging

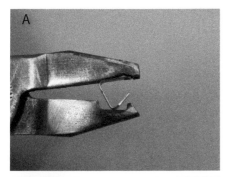

Ear tagging is also a long-term identification method that can be performed quickly with little pain or distress. The first picture in this series shows the proper placement of the identification tag in the applicator (A). The next two pictures demonstrate placing the tag in the lower one-third of the ear of the mouse (B) and then securing the tag in the ear (C). Place the tag on the lower portion of the ear to prevent the ear from folding over. Notice that the tag is positioned with the number facing forward, allowing it to be viewed easily. Care must be taken when placing the tags. Tags could be torn out if placed too close to the edge of the ear

Note: Although this handout uses the mouse for demonstration purposes, the techniques used can be easily applied to rats. The only difference is in the manner of restraint.

appendix a: rat (Rattus norvegicus)

Genus and species: *Rattus norvegicus;* Order: Rodentia

About 95% of research animals are rodents, with 90% being rats and mice.

The laboratory rat was the first animal in which the primary reason for domestication was for use in scientific research.

Some characteristics that make the rat a good model for research are as follows:

- They are easy to care for and handle.
- Small size makes them easy to maintain.
- The have a high reproductive capacity.
- The have a short reproduction time.
- An abundance of baseline information already exists.
- Large numbers of strains are available, with natural genetic deficiencies that can act as models for human; the rat has all the same organs as humans, and these function similarly.
- There are germ-free and pathogen-free production techniques.

Common inbred strains include
- Fisher 344
- Brown Norway

- Lewis
- Wistar-Furth

Common outbred strains include

- Sprague Dawley
- Wistar
- Long Evans (hooded)

physical characteristics/anatomical features[1]

Some general features for rats are as follows:

- **Acute hearing**: Makes them sensitive to ultrasounds and high-pitched sounds
- **Well-developed sense of smell**
- **Cannot vomit**
- **Rapid resting heart rate** (300+ per minute) and **respiratory rate** (100+ per minute)
- **Different vision**: See in pastels and are blind to long-wave (red) light
- **Open-rooted incisors**: Grow continuously throughout life and have enamel only on the anterior surface, which results in continuous sharpening as the incisors grind against each other as the rat gnaws
- **Open inguinal canal (males)**: Also have an os penis
- **Thermoregulation**: Tail is the principal organ that regulates body temperature
- **Nocturnal**: Most active at night
- Normally sit on all four limbs but will stand on hind legs to investigate
- Healthy rats keep themselves well groomed and clean
- Can become accustomed to gentle handling but can bite if improperly handled or frightened
- **Group housed**: Are social animals and do well housed together; same sex can be housed together

---

[1] See appendix b, normative data for the laboratory rat.

Anatomically, rats are similar to mice except for the following:

- Rats continue to grow throughout their lives.
- Rats do not have a gall bladder.
- Mammary tissue extends dorsally in the rat but not as far as in the mouse. Thus, most rat mammary tumors are ventrally located.
- The hair coat of the rat is not as smooth and glossy as a mouse.

sexing and breeding

- **Anogenital distance**: Used to determine male and female. The distance between the anus and genital papilla is twice as long in the male as in the female.
- **Mammary glands**: Female mammary glands are visible, while the males' are not.
- **Sexual maturity**: Sexual maturity occurs in the female rat at 37–67 days and in the male slightly later. It is best to breed rats initially within 72–90 days of age.
- **Provide breeder chow** for the pregnant and nursing females.
- **Gestation period**: Gestation is 21–23 days.
- **Postpartum estrous**: Rats come into "heat" within 24 hours of giving birth.
- Pups should be weaned at approximately 21 days.
- **12/12 light cycles** (12 hours dark and 12 hours light): This is important in breeding rats. Deviations from this cycle can affect reproductive performance.
- Female rats are **polyestrous**, with an estrus cycle of 4–6 days.
- Rats may be bred in a monogamous or harem breeding system.
- **Pups open their eyes** between 10 and 14 days of age.
- **Pups ears open** between 12 and 14 days of age.
- **Pups begin to get hair** between 8 and 9 days of age.
- Do not disrupt or clean cages for several days after pups are born.

sources

Some animal care departments will do all the ordering for you.
Some of the more common commercial vendors are

Charles River Laboratories
Wilmington, MA 01887
1-800-522-7287
http://www.criver.com
Established strains,
transgenics

Harlan Sprague Dawley / Harlan Teklad
Indianapolis, IN 46229, Madison, WI 53744
1-800-793-7287, 1-608-277-2070
http://www.harlan.com
Established strains, transgenics,
laboratory animal diets

Jackson Laboratories
Bar Harbor, ME 04609
1-800-422-6423
http://jaxmice.jax.org/index.html
Established strains, mutant stocks

Taconic Farms
Germantown, NY 12526
1-888-822-6642
http://www.taconic.com
Established strains, transgenics

Noncommercial options:[2]

- Other universities
- Other researchers on campus

husbandry

Work quietly around the rats; they are very sensitive to sharp,
loud noises.

There are various types of caging (male rats may be housed
together):

Shoebox cages: Most rats are housed in this type.

Wire-bottomed cages: These are not allowed unless it is a
research necessity to collect feces or urine or to prevent contact with the bedding.

Microisolator cages: These are used to house rats to prevent
them from acquiring rodent pathogens. These animals are
housed in a sterile environment and provided sterile food
and water. This is especially important for immunocompromised animals.

[2] Make sure to discuss options with your veterinarian before obtaining animals.

- An assortment of commercial bedding is available.
- Caretakers generally clean cages twice a week, more often if necessary; this includes changing food and water.
- Rooms should be cleaned and disinfected when emptied, or as often as needed, to keep them free of dirt, odor, and contamination.
- If cages have automatic watering systems, make sure the cages are pushed back all the way onto the rack to ensure that the water valve reaches into the cage and is accessible to the animal.
- Make sure that the water valve is not plugged or leaking.
- When moving cages, it is important to remove or turn the water bottle around to prevent spilling water into the cage.

diet

- Generally, rats are fed and watered ad libitum. Laboratory animal feed manufacturers produce a variety of nutritionally complete diets. Supplementing these diets is usually unnecessary.
- Diets are normally provided in 4- to 5-g pellets. Normal adult **food intake is 5 g per 100 g body weight** per day. Pellets are firm and require gnawing, which helps keep the incisors worn down.
- Nonpelleted meals or powder diets can be used when food intake is being monitored or when experimental substances are being added to the diet. **If feeding this diet, remember to provide something for the rat to chew or gnaw to prevent malocclusion.**
- Water can be provided by water bottles or by an automatic watering system.
- Depending on the cage design, food pellets can be placed in the feeders or in the designated area in the wire cage top. They should be filled with enough food to last several days. It is not necessary to fill food pellets to the top of the wire top. This can be wasteful.

identification

- **Cage cards** should be easy to read and identify. They must include strain, sex, identification number, investigator, protocol number, and so on.
- **Cage alerts** can alert to special needs (e.g., sickness, pregnancy, weaning, etc.).
- **Tape on cages is discouraged.**
- **Ear notch**: Animals do not need to be anesthetized; removed tissue may be used for polymerase chain reaction.
- **Ear tag**: Animals do not need to be anesthetized.
- **Tattoo**: Animals *must* be anesthetized.

stress management and enrichment

- **Enrichment**: Nylon bones, ping-pong balls, PVC tubing, nestlets can be used. (Do not use cotton.)
- Try to service the rooms at the same time each day.
- Be quiet.

recognizing pain and distress

Most animals will try not to show pain or illness until they are quite sick. Most prey animals turn into victims if they show signs of weakness. The following are some common signs of pain or distress:

- **Abnormal biting**: Does the animal chew or self-mutilate?
- **Aggressive**: Does the animal try to bite when touched or disturbed?
- **Anxious**: Does the animal seem agitated or restless?
- **Change in grooming**: Does the animal lack normal grooming behavior? Does the hair coat of the animal appear dull, ungroomed, and oily? Are there signs of hair loss?
- **Weight loss/dehydration**: Is there a decrease in food or water consumption?

- **Secretions/excretions**: Does there appear to be abnormal amounts of porphyrin, tears, feces, urine?
- **Mammary tumors**: Are tumors interfering with normal behaviors?
- **Malocclusion**: Do the teeth of the animal line up, or are they growing in all different directions?
- **Coordination**: Is the animal unsteady, wobbly, or have a head tilt?
- **Depressed**: Does the animal seem lethargic or does not care what you do to it?
- **Movement**: Is the animal difficult to rouse, or does it seem restless?
- **Posture**: Is the animal hunched over or arching its back? Does it appear to have difficulty walking?
- **Teeth grinding**: Can you hear excessive gnashing of teeth?
- **Vocalization**: Does the animal cry out, whimper?

common diseases and prevention

Diseases

- **Sendai**: This virus suppresses the immune system, often causing the animal to develop secondary bacterial infections.
- **MRM** (murine respiratory mycoplasmosis): MRM may cause severe ear and lung infections.
- **SDAV** (sialodacryoadenitis virus): Animals with an active infection often reduce feed intake because of salivary gland swelling.
- **Pinworms**: These can cause severe itching and rectal prolapse and are difficult to eradicate.

Prevention

- **Microisolator lids on cages**: Open the cages only in a hood.
- **Use serology.**
- **Use personal protective equipment (PPE).**
- **Cage changes** should be performed under hood.
- **Limit exposure.**

record keeping

Records should be kept for all species, be available within close proximity to the animals, and be easily accessible to all lab personnel.

- Keep accurate, up-to-date records of everything that happens in a laboratory setting. This information helps facility managers and investigators determine whether procedures are being followed according to established standards.
- **Husbandry records**, such as an **animal room log sheet**, can be used to identify the animal strains or species being kept in the room, where the animal came from, animal census, room temperature and humidity, cage changes, food and water changes, and whether racks were washed, floors disinfected, or any other husbandry tasks accomplished.
- **Standard operating procedures (SOPs)** are documents that state how a procedure is to be performed. These procedures are written to allow different people to perform the same tasks in the same way at any time. SOPs should be kept in a central location where technicians and investigators have easy access to them.
- **Clinical records** (experimental and surgical records) should be maintained on all species. These records should indicate species, sex, date of birth, health status, vaccinations, surgical procedures, postoperative care, and any other pertinent information.
- The National Institutes of Health (NIH) requires that all assurance records related directly to grant applications, research proposals, and changes of research activities be maintained for at least 3 years after completion of an activity.
- If an inspector asks for your records and you cannot provide a written copy regarding what you have done, the inspector must assume that it was not done. **If it is not written down, it never happened.**

protocols

Every research lab associated with the University of Wisconsin–Madison campus must have an approved animal care and use protocol to perform research on live animals. With that in mind,

Read your protocol. The protocol contains the detailed information regarding the approved techniques and procedures you will be performing.

---

euthanasia

You must follow the approved euthanasia regimen that is listed in your protocol. The following are **common euthanasia methods**:

- Inhalant anesthetics
- CO, CO_2
- Barbiturate overdose
- Cervical dislocation only if the rat is under 200 g and must be anesthetized
- Decapitation

There are other acceptable methods of euthanasia that may be more useful to your research. Contact your veterinarian for suggestions or go to the following link from the American Veterinary Medical Association (AVMA) Panel of Euthanasia: http://www.avma.org/resources/euthanasia.pdf.

appendix b: normative data for the laboratory rat

general information

Adult weight
    Male   450–520 g
    Female   250–300 g

Adult weight	
Male	450–520 g
Female	250–300 g
Surface area	Weight $(g)^{2/3} \times (9) \times 10^4$
Body temperature	35.9–37.5°C
Food consumption	5–6g/100 g body weight/day
Water consumption	10–12 ml/100 g body weight/day

blood and oxygen

Pressure	
Systolic	88–184 mm Hg
Diastolic	58–145 mm Hg
Volume	
Plasma	4.15 ml/100 g ody weight
Whole blood	57.5–69.9 ml/kg
Heart rate	250–450 beats per minute
Tidal volume	0.6–2.0 ml
Minute volume	75–130 ml/min
Stroke volume	0.9 µl/beat
Plasma	
pH	7.41
CO_2	28 cc/100g
Respiration rate	70–115 breaths per minute
O_2 consumption	0.84 ml O_2/g/h

experimental information

Maximum single bleed[1]	10% total blood volume
Gavage volume	10 ml/kg

hematology

Leukocyte count	$3-17 \times 10^3/\mu l$
Total	
Neutrophils	13–26%
Lymphocytes	65–83%
Monocytes	0–4%
Red blood cells	$5-10 \times 10^6/mm^3$
Hemoglobin	11–19 g/dl
Platelets	$200-1,500 \times 10^3/\mu l$
Packed cell volume	35–37%

breeding

Breeding onset	72–90 days
Pseudopregnancy	10–13 days
Fertilization	2 h postmating
Implantation	4–5 days
Gestation	22–23 days
Postpartum estrus	24 h
Puberty	
Male	37–67 days
Female	37–67 days
Breeding life	12–18 months
Litter size	8–14
Birth weight	5–6 g
Eyes open	10–14 days
Wean	21 days

Recommended blood drawing techniques for the laboratory rat include use of the jugular, saphenous, or pedal veins, tail vein/artery, and cardiac puncture.

For further information, please contact your lab animal veterinarian.

---

[1] Example for a 250-g rat: Total blood volume (TBV) is 64 ml/kg $\times$ 0.25 kg = 16 ml.

appendix c: blood volume

Although additional percentages are given in this appendix, the trainers at the Research Animal Resources Center (RARC) recommend using 10% of the total blood volume as the maximum amount that can be collected at one time. To allow for an adequate recovery period, we further recommend that the maximum amount be withdrawn only once every 2 weeks (based on 10%). Daily samples can be collected as long as the cumulative blood volume does not exceed the calculated maximum over a 7-day period.

Circulating blood volume in the rat:[1]

Species	Blood Volume (ml/kg)	
	Recommended Mean	Range of Mean
Rat	64	58–70

Blood volume based on a 250-g rat:
$64 \text{ ml/kg} \times 0.25 \text{ kg} = 16 \text{ ml}$

Total blood volume and recommended maximum blood sample volume:[1]

Species	Blood Volume (ml)	7.5% (ml)	10% (ml)	15% (ml)
Rat (250 g)	16	1.2	1.6	2.4

---

[1] EFPIA/ECVAM paper on good practice in administration of substances and removal of blood. *J Appl Toxicol* 21: 15–23, 2001.

Rat blood collection guidelines by percentage of total blood volume:[2]

weekly	7.5%
every 2 weeks	10%
every 4 weeks	15%

[2] American Association for Laboratory Animal Science. *Laboratory Mouse Handbook.* Memphis, TN: American Association for Laboratory Animal Science, 2006.